W9-ATA-380

Remanufacturing

The Ultimate Form of Recycling

Rolf Steinhilper

Fraunhofer IRB Verlag

This book has been written with the support of
Remanufacturing Industries Council International
Automotive Parts Rebuilders Association
Fraunhofer Demonstration Center Product Cycles

The inner title photo shows
the assembly area
of Precision Alternators and Starters
Fairfax, VA

© 1998 All rights reserved with
Fraunhofer IRB Verlag, D-70569 Stuttgart

ISBN 3-8167-5216-0

Illustrations: See Acknowledgements
Typesetting: DigitalStudio H. Möhwald, Sindelfingen
Printing and Binding: Druckerei Hoffmann, Inh. M. Wetzstein,
Kornwestheim
Printed in Germany

Contents

Foreword by the Author

Remanufacturing – the ultimate form of recycling: My writing this book was inspired and supported by great men. Without their encouragement and critique I doubt it would have yet been begun.

First I want to thank my father Ulrich Steinhilper who, when I was only six years of age, not only taught me to ride my first bicycle, but also to "recycle" the components – I learned how to patch the tire inner-tube! Out of this my first practical experience a gradually expanding range of skills developed, which, by the time I was beginning my studies as a university student, equipped me to repair cars. At one and the same time I had my first contact with remanufactured products like brake shoes, clutch assemblies and alternators.

Moving from my education to the world of work, I want to record my thanks to Professor Hans-Jürgen Warnecke, President of the Fraunhofer Institutes for Applied Research. Twenty years ago he had the foresight to support my proposal to start research and to offer consultancy in the field of remanufacturing. This has led to the service we provide today in five Fraunhofer-Centers "Product Cycles" established in the key industrial areas of Germany.

At an international level, sincere thanks go to Professor Robert T. Lund from Boston University, as well as to Mr. William C. Gager who gives his introduction on the next page. They invited me to share in their development of remanufacturing on an international scale, both in science and in industry. Thus in 1982 , my first visits to M.I.T. established a really friendly and open relationship with experts from all points of the globe.

Closer to home, in Europe, I want to thank Fernand J. Weiland for his support and for the preface overleaf. My heartfelt thanks go to my colleagues at the Fraunhofer Institute for their help and support, to the companies who waived copyright or supplied the illustrations and graphics for the book, and to Mrs. Susanne Bacher who, as always, has supplied the illustrations and graphics.

Last, but by no means least, I must thank the readers of this book for their interest in the further advancement of the great concept of remanufacturing.

Rolf Steinhilper
Head of the Fraunhofer Centers "Product Cycles",
Fraunhofer Institute IPA Stuttgart, Germany - 1998

Introductory Remarks
by William C. Gager

Remanufacturing and rebuilding is a process that has been around for over 60 years to restore old products to like new performance and save energy, natural resources, landfill space and reduce air pollution by less re-smelting. This industry also creates hundreds of thousands of jobs and new tax-paying businesses.

By extending product life and by giving products numerous lives, remanufacturing saves 85 % of the energy that went in to manufacturing the product the first time. A product can always be resmelted but if we can add numerous lives to that product before it gets resmelted we have really helped save our environment and benefit society.

Public policy makers and scientists are becoming more concerned about the consequences of global warming and sustainable development. We all know that there is only a certain amount of natural resources on this planet. The closer we get to "zero waste", the more future generations will enjoy the same material wealth that we enjoy today. The days of throw-away products and single use products are over with. It's time to move forward with more and more remanufacturing.

William C. Gager
President, Automotive Parts Rebuilders Association
Chairman, Remanufacturing Industries Council International

Preface
by Fernand J. Weiland

I am very grateful to Dr. Steinhilper for writing this very interesting book. The image of the European remanufacturing industry will be greatly improved as a result of him acknowledging and communicating the important contribution we are making to the environment, economies and the labour markets in Europe. Because of these important values this young industry, which is growing fast, deserves enhanced support and better recognition from the media, politicians and all persons concerned with a cleaner environment.

Fernand J. Weiland
Director of the European Division
of Automotive Parts Rebuilders Association

Remanufacturing: Rebuilding the Future

What is Remanufacturing?

Remanufacturing is recycling by manufacturing "good as new" products from used products.

Remanufacturing has many names. Rebuilding, refurbishing, reconditioning, overhauling are also frequently used terms. Increasingly, however, remanufacturing is becoming the standard term for the process of restoring used durable products to a "like new" condition.

Remanufacturing the Standard Term

Figure 1: Remanufacturing is the Ultimate Form of Recycling.

It Receives and Deserves Support from More and More Organisations

Remanufacturing also has many meanings. Remanufacturing involves a broad scope of participants in modern products' life cycles. Each participant has his special focus.
There are various significant characteristics and effects:

Characteristics and Effects

From an *environmentally conscious citizen's* viewpoint, the main reason for being interested in remanufacturing will certainly be their appreciation for recycling as a key principle of "making peace with nature" and securing a sustainable future.

New Business Opportunities

A *business strategist* might discover that remanufacturing rewards the world of manufacturing with new business opportunities in the after sales service market enabling one to offer their customers new solutions with a minimum total cost of ownership.

Turn Around Cost into Profits

The *waste manager* will be delighted how remanufacturing can serve to turn around their costly disposal processes into product loops creating profits.

New Challenges

The *innovative manufacturing engineer* will identify the five steps of remanufacturing, from disassembly, cleaning, inspection via parts reconditioning until reassembly and final testing. He will see them as an expansion of the technologies he is familiar with, accompanied by new challenges especially in the first steps.

Any *maintenance expert* would certainly point out, that remanufacturing is the most effective way to perform servicing and repairing tasks to both the worker's and customer's satisfaction.

Creating Jobs

Politicians and *government officials,* whether on federal, state or regional levels, will agree that remanufacturing is a unique strategy for new business development and creating jobs in their surrounding communities.

Economists will honour remanufacturing enterprises as members of the esteemed community of "hidden champions" of industry, playing an important role not only for today's but also future industries' survival.

The *environmentally responsible industrial professional* at once will recognize that remanufacturing is the key link in an integrated green technology chain to perform his new "cradle to grave"-product and recycling responsibilities and liabilities successfully.

Integrated Green Technology Chain

For the *consumer*, remanufacturing is not just the most ecologic but also the most economic way of having access to state-of-the-art technology products at affordable prices but always the up-to-date quality of new products.

Affordable Prices

Success and also *risk assessment analysts* will describe remanufacturing as an approach to successfully repeat, replay (and to an interesting extent even replace) those manufacturing technologies, which so far only created a product's first life cycle. Remanufacturing offers a product many life cycles – it is the enabling technology for "cradle to cradle" product loops.

Scientists will plainly state, that remanufacturing is the most efficient and effective way to save resources, whether energy or materials, of any form of recycling.

Sustainability

Finally, you don't need to belong to the small party of *idealists* to esteem remanufacturing not just as a principle for the "rebirth" of a product, but also as the birth of an idea how to give new breath to the ideals of a sustainable future for our planet and civilizations.

Benefits in Practice

To a certain extent this full dozen of various views of remanufacturing – in number and in sequence – already reflects key messages of the twelve chapters of this book. More important of course they also in practice offer plenty of benefits and growth potential to industry and society.

Therefore it is no surprise that in today's world one can already discover an impressive range of application examples representing remanufacturing as the ultimate form of recycling.

Partnering the Uprise of Industries

Remanufacturing has its roots and displays a long tradition already since the very first moves and outcomes of the industrial age like steam engines, railways, power generation and electrical equipment, machine tools etc. Furthermore remanufacturing always has been a strong partner throughout the uprise of the automotive sector, which undoubtedly was the main industrial driving force so far. Remanufacturing will continuously strengthen its important role in the automotive industry.

Figure 2:
Remanufacturing
Reaching from
Traditional
to Future
Products

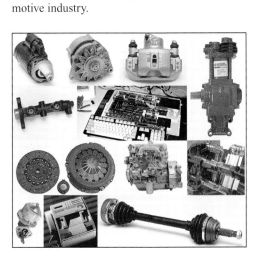

Looking ahead, remanufacturing will certainly also keep up with the most up to date technologies and products, offering new opportunities and incentives also within the fast moving electronic industries with their computers, communication and multimedia products.

Keeping up with Future Products

It is a surprise, however, how little the good news about remanufacturing has made its way into the awareness of consumers and professsionals so far, or how limited their influence to decision making in industry or society still appears to be. It seems, however, that just recently there has been an awakening no longer to leave aside the chances of remanufacturing as a brilliant idea for a proactive integration of technology and environment, serving both new markets as well as a responsibility for the nature and the future of our society.

Awakening and Awareness to Come

Product Retirement: "You Only Live Twice"?

Industrial production and productivity, especially in the automotive sector, has made impressive progress since the early days, when engineers Daimler and Benz put their very first motor car onto the cobble stone roads a century ago. Henry Ford rolled out his first Model T in 1908 and manufactured the first ten million of them – the main portion of those days' world car market – in the sixteen following years. Who would have expected, that such a production volume nowadays is just a matter of ten weeks worldwide – with products offering a multitude of functions and features compared to Tin Lizzy?

Figure 3:
Henry Ford
with
10millionth
Model T

While looking back, the quantities of products entering the market have certainly been difficult to foresee, but it is an easy task looking ahead to forecast the product volumes returning from their markets, as cars retire after an average lifespan of around ten to twelve years.

There are 170 million cars on America's roads, 150 million cars in Europe, and around the same quantity all over Asia. For an attempt to imagine them lining up when retiring from their useful lives within the next ten years, the equator line of our globe is not long enough.

A 150 million car queue would circumspan our planet approximately ten times, or for a better imagination, in fact reaches from the earth to the moon, as an artist's impression illustrates.

Reaching the Moon with Scrap Cars?

*Figure 4:
150 Millions
of Scrap
Cars
Illustrated*

Even faster revolutions and innovations have speeded up the electronic sector. During the first semiconductor experiments in Silicon Valley who would have forecasted, that the personal

Overtaking Car Production

**The New
Challenge
of
Electronic
Waste**

computer industries will need just a few decades to reach and then to exceed the car production quantities in numbers (now 60 millions of PCs / 50 millions of cars per year) and that on top of the computer industry as such also today's cars contain several computers for electronic engine, safety, comfort control and more?

Meanwhile, the amount of electronic products returning from the market has just started to demonstrate its dimensions. It will certainly not line up like the cars, but it already piles up to hills soon reaching fearsome mountain sizes. In every big city the illustrated pile of waste electronics and monitors, whether from PCs or TVs together with other timed out electronics like telephones, fax machines, printers etc., which counts in hundreds of thousands annually, would make any skyscraper look like a toy.

*Figure 5:
Any Big City's
Annual
Office Electronics
Scrap
Illustrated*

**From
One-Way
to
Cyclical
Systems**

Unfortunately, however, the slogan "you only live twice" has up to now only become common as a famous movie title. High-tech products only live once – but it is in fact becoming common, that the traditional one-way-system from the manufacturer to the market has to be replaced by a cyclical system.

Industry's Responsibilities

In Europe and in Asia laws are becoming effective, that put the responsibility for the full life cycle of a product, including their take back and recycling, completely onto the manufacturer's shoulders.

Bearing this responsibility has meant considerable efforts for building up completely new collection and recycling networks already in the packaging products sector and meant a heavy workload for all parties involved.

First Experiences

In the high-tech products sector, it is strongly recommended not just to use one's shoulders and muscles but in particular one's head and brain to develop an overall concept of product responsibility all across manufacturing, product use and retirement / recycling.

Intelligent Solutions

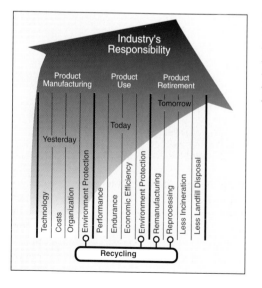

Figure 6: Today's Industry's Responsibility throughout the Product Life Cycle

Established Recycling Processes

**From
Existing
Technologies
towards
Innovations**

When manufacturers started to study adequate product recycling processes, of course the first approach was to rely on existing technologies. This can be compared with the approach of the early car makers, who met the challenge of a manufacturing technology for the first time (the very first task in the bottom left of the previous figure) one century ago. As the technology of manufacturing horse coaches existed, during the first years most cars really rolled out as motor coaches – but the experience was, that they were not manufacturable in large quantities at affordable prices. New developments and innovations became necessary and had to take place further on, which in fact has happened.

**Scrap
Processing
not
Sufficient
any Longer**

One hundred years later, the same applies to attempts only to use the existing scrap processing technologies for product recycling, whether cars, appliances, computers or other high-tech waste is being considered.

The classic procedure is no longer affordable, in particular since it is no longer true, that retired products are mainly a source of recovering metal.

*Figure 7:
Regarding
Retired
Products
just as
Metal Scrap*

Regarded mainly as metal scrap, retired cars over decades have primarily put into a powerful, up to 5000 horse power shredder, which cuts them into small pieces within a few seconds.

However, after the shredder process, only the metals can be recycled, that means resmelted together with new metal from ore, in the mining and steel industries.

Figures 8 and 9:
Input and
Output
of a
Car Shredder

For this purpose, however, a magnetic separator is necessary, which concentrates the steel and iron fraction which then can be shipped to the blast furnaces of the steel industry.

Figure 10:
Metal Fraction
from
Shreddered
Products

**Economic
and
Environmental
Problems**

This fraction and also the "rest" (that is up to one third of the original product weight), when cars are being recycled (and more than 50 % of the original product weight, when electronic equipment is being recycled) provide both economic and environmental problems.

At first, they lead to high working cost for a manual sorting out of unwanted materials and components which arrive together with the magnetically separated steel and iron fraction – such as rubber parts from steel-belt-tires, copper windings around electric motor iron rotors, plastics with metal inserts etc.

*Figure 11:
Manual Sorting out
of Unwanted
Materials*

Furthermore, and worse, the mixed non-metal fraction of dirt, dust, rubber, glass, plastics, foams, textiles, stuck with engine and gearbox oils, toxic braking liquids etc. (recently also toxic chemicals from airbag gas generators), which reach one third of the original products's weight and volume (or of the queue from the earth to the moon, remember the comparison illustrated earlier), wait for further treatment.

Figure 12:
"Shredder Fluff":
Plastics,
Rubber,
Textiles,
Dirt,
Dust, and
More

Until recently, this so-called "shredder-fluff" had been brought to landfills together with more or less harmless household waste – endangering and intoxicating the groundwater and the atmosphere – a procedure, which is now forbidden by law in Germany and may be soon in other countries too.

Landfill Disposal Forbidden

Environmentally responsible product recycling today and in future therefore starts with careful preparation and disassembly processes to recover most of the mentioned fractions, valuable parts or materials and / or hazardous components and substances, so that they will not end up in a mixed toxic and non recyclable fraction.

To enforce this move towards cleaner technologies, the automotive, appliance and electronic industries are becoming obliged by law or are preparing themselves to take back and recycle their products. In Germany, the automotive and the electronic industries recently have declared voluntary self commitments to do this.

Voluntary Self Commitments

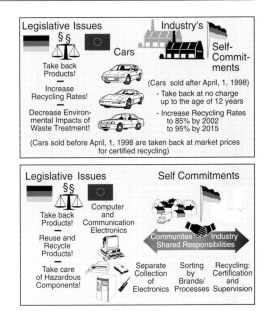

The most significant progress during this efforts will be achieved, if the tracks of the established scrap recycling technologies are left and new innovative approaches and processes for future product recycling solutions are developed.

Cleaner Technologies

In addition to thousands of remanufacturing companies, who already disassemble products, hundreds of disassembly workshops have started or will soon start operation in all industrialized countries.

The pioneers can be located in the automotive, appliance and electronics industries as well.

*Figure 15:
Disassembly Line
for Cars*

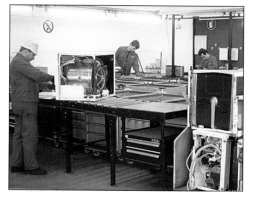

*Figure 16:
Disassembly Workshop
for Household
Appliances*

*Figure 17:
Disassembly Line
for Television Sets*

Figure 18:
Disassembly Line for
Personal Computers

In the disassembly lines or workshops, four groups of parts are recovered from the retired products, before their remainings can then be passed to the shredder process.

Manual Separation of Valuable and non Recyclable Materials

Group 1: Parts of valuable materials, for example from cars:
- lead batteries
- brass radiators
- aluminium wheels

from electrics and electronics:
- gold contacts
- copper wiring etc

Group 2: Parts made of (mainly non metal) materials, which otherwise would end in the non recyclable shredder fluff, for example from cars:
- rubber tires
- plastic bumpers
- windshield and window glass

from electrics and electronics:
- plastic housings, keyboards etc
- insulation materials etc

Group 3: Parts containing hazardous substances, for example from cars:

- oils,
- transmission fluid,
- brake fluid,
- air conditioning refrigerants,
- airbag chemicals

and

from electrics and electronics:

- mercury switches
- cadmium or lithium batteries,
- circuit boards and plastic parts with flame retarding substances,
- polychlorid biphenyle or electrolyt containing capacitors,
- liquid crystal displays,
- lead and barium glass cathode ray tubes

Manual Separation of Hazardous Components

Figure 19: TV and PC Tubes

These parts and substances have to be carefully separated for further special treatment, as they otherwise would not only constrain recycling possibilities, but even intoxicate both the recyclable and the non recyclable materials.

Careful Separation

Group 4: Reuseable Parts, both for direct reuse in second hand product markets or as used spare parts, as well as for remanufacturing purposes,

**Separation
of the
Most
Interesting
Reusable
Parts**

which is the main focus of this book. Examples
from various product origins and industries will
be shown in the forthcoming chapters. Already
at this point, however, it should be pointed out,
that reuseable or remanufacturable parts are the
main source of creating profits under the bottom
line of the overall product recycling effort, as
any cost breakdown of recycling clearly shows.

This is an important fact, although the concept
of remanufacturing in many cases is still at the
very beginning of its wide range of applications.

Cost Breakdown

**Disassembly
Costs
Are not
the Only
Effort to
Prepare
Recycling**

Product disassembly is a labor cost intensive
process, which has to be equalled by adequate
returns from the various fractions of materials or
parts recovered from the product.

So far however only the metals can be recycled
profitably.

Other materials, for example most plastics, need
additional costly processes like cleaning, sorting, compounding which in the end often makes
the recycled material even more expensive than
new plastic/materials from crude oil.

*Figure 20:
Rubber Parts from the
One Shift Output of a
Car Disassembly
Plant*

Some materials, especially rubber from tires and seals, cannot be recycled as rubber so far. They are burned or transferred to waste dumping sites, where they soon grow to artificial mountain sizes.

Furthermore the disassembled components and separated liquids containing hazardous substances have to be transferred to hazardous waste incineration or special treatment plants.

Returns from Reusable and Remanufacturable Parts

So, besides from the valuable materials like copper and the also profitably recyclable materials like steel and iron, where are the returns from recycling retired products?
To break even with the cost of disassembly as well as with the cost of responsibly taking care of the materials which are difficult to recycle, a considerable contribution – in many cases the only solution – to the recycling cost problem comes from reuseable and remanufacturable parts.

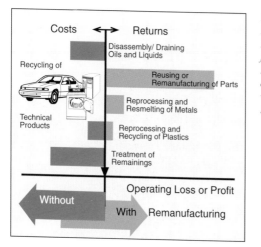

Figure 21: Profits or Losses from the Recycling of Products – with or without Parts Reuse

**Turning Around
Problems
into Chances**

Anybody can imagine, that the problem of rubber recycling from tires at once disappears and turns around into a positive situation, when the tires can be reused through retreading. The same applies to reusing carbody parts like hoods or doors for accident car repairing instead of just recycling them as bulky scrap.

Reusing parts saves their inherent engineering and manufacturing values – either their actual

*Figure 22:
Car Hoods and Doors
Ready for Reuse*

value as used parts or their like new value by adding the necessary value through remanufacturing them to a "good as new" condition.

**Remanufacturing,
Repairing and
Manufacturing
are a Community**

Whether on the product level as remanufactured products themselves, or on the part and subassembly level as remanufactured units for repairing other products, either a second product life cycle is created or an original product life cycle is extended. Both results without the concept of remanufacturing would otherwise require the whole effort of new manufacturing.

Together however, remanufacturing, repairing and manufacturing form an ideal community of different professional and industrial activities.

Manufacturing, Repairing and Remanufacturing: Sympathetic Neighbors

There are two main streams and facts that keep our manufacturing industries moving as well as repairing and remanufacturing operations going:

Keeping Products Going

Continuous technological progress, inventions and new developments regularly lead to innovative products and further demands ("technology push" and "market pull").
Besides, and may be more important in a product's every day life, even the most carefully engineered and manufactured product can not last for ever.

Repairing or Remanufacturing?

So when a product becomes defective, that means too worn out or damaged to perform properly, and the question arises how to put it back into service, repairing or remanufacturing will immediately be considered. At first glance, both processes will achieve the same important result: They equally avoid the expenses of buying a new product as well as of disposing of the old one.

Keeping Products Operating

Looking into the different process steps and into the characteristics of the products coming out of the processes, it becomes clear, that in fact there are some significant differences regarding overall quality and warranty accompanying a repaired or a remanufactured product.

Comparing the Processes

Figure 23:
Repairing –
Remanufacturing –
Differences of
Processes and Product
Characteristics

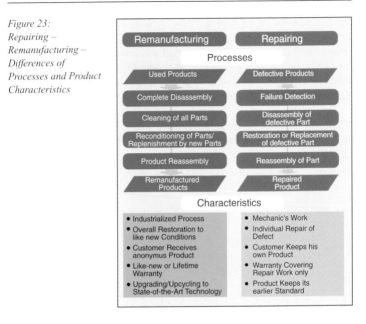

Maintenance and Recycling

First or Second Life Cycle

So, generally speaking, repairing more or less extends the useful life of a product while remanufacturing establishes its next full life cycle. Regarding ownership, a repaired product will return to its original owner, while remanufactured products are anonymous and ownerless like new products before they enter their customer's ownership.

Looking for the Borders

With such descriptions and approaches, engineers as well as lawyers try to define the borders between repairing and remanufacturing or, more generally speaking, to keep apart the processes of maintenance (which is superior to repairing) and recycling (which summarizes different processes including remanufacturing).

To a certain extent of course, these definitions are correct. But there are always some cases, where for example carefully repaired products equal the quality of remanufactured products or where remanufactured products are overhauled for the original owner, etc.

From Border Lines to Junction Lines

Therefore it should be recommended, not to try to separate maintenance and recycling from each other, but to understand them as close "relatives" or at least sympathetic "neighbors" with the process of repairing as their common border or junction line.

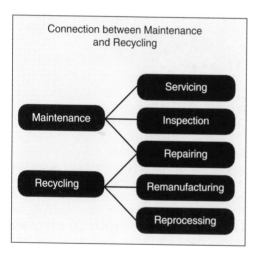

Figure 24: Maintenance and Recycling Belong Together

This connection between maintenance and recycling has in the meantime also become accepted by the relevant national and international standardization organisations.
Moving from theory to practice, not only the definitions, but more important the events of repairing and remanufacturing can be watched close together or even depend on each other:

Repairing and Remanufacturing

To give an example, if the maintenance specialist, who has opened the engine compartment of the fork lift truck in the photo below finds out, that servicing the engine by refilling oil will not help any longer, but that the engine has to be replaced, he will certainly repair the fork lift truck with a remanufactured engine.

Figure 25:
Repairing the Product
(a Fork Lift Truck)
Can Rely on
Remanufacturing one
of its Subassemblies
(an Engine)

Theoretically, also an individual repair of the engine would be possible. However, due to the high labor costs involved and a lack of specialized tools and overhauling equipment on site, such an approach will not be successful in practice. So the fork lift truck receives a remanufactured engine.

Millions of Remanufactured Engines

On all our roads, of course not just fork lift trucks, but many many cars, and that means millions of them, just keep on rolling because they are repaired (maintained) by remanufactured (recycled) engines and parts!

Strong Links

Cooperation or Competition?

In the same way, as repairing and remanfacturing rely on each other, also manufacturing and remanufacturing benefit from each other.

At first glance, however, manufacturers some-times see remanufacturers as some kind of unpleasant competition, because as earlier stated, remanufacturing a unit can avoid manufacturing a new one. That's why manufacturers might be afraid of loosing some business, at least if they are not remanufacturers themselves. But this is a short sighted view, because certainly no product can be remanufactured an unlimited number of cycles, and will have to be replaced by a newly manufactured product one day.

Already in each remanufacturing cycle, manufacturing is a winner too. There are several strong reasons for that:
When products of a current model generation are concerned, manufacturing of parts supply for the remanufacturing of these products provides some additional, that means very welcome production volume for the manufacturer's existing production equipment.

Manufacturing is a Winner of Remanufacturing

When it comes to products or subassemblies of previous product generations, who have run out of new production but still run in large quantities in the market and require maintenance, many manufacturers experience it as a great relief, that remanufacturing can now completely take care of this part of the after sales service, so that the manufacturers are not obliged to keep and run their production equipment just for some spare parts production for older models. Notably the remanufacturers are in this case able to replace most of their spare part requirements by sourcing from meanwhile enough old products. The old model generation now retires completely from their useful lives, and contain an interesting share of still reconditionable parts for remanufacturing.

Relief From Old Models' Tasks

Adding Value to Each Other

This shows, that even up to the very end of the life cycle chain, from manufacturing via maintenance until final recycling, the four processes dealt with in this chapter like parts production, product repair, units remanufacturing, and parts recovery from obsolete products are not just connected by obvious streams of hardware, but they also add value to each other in a more virtual way like transferring and sharing technological know-how between manufacturing, repairing and remanufacturing, exchanging experiences about product failure reasons among them and with the design engineers etc.

A Well Balanced Win-Win Situation

Overall, this is a well balanced win-win situation for all market players involved, last but not least of course for the consumer who also considerably benefits from this profitable interaction of the parties mentioned.

Figure 26:
Interaction Between
Life Cycle Participants
and Processes

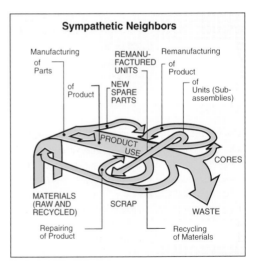

Core Supply

Cores, this is the international term for used products or old units which are the main input or raw material of the remanufacturing industry. They are the beginning of a remanufactured product's life-cycle – and its end as well, if the product can be remanufactured again.

The Beginning and the End

That means, if a car owner during a trip suddenly notices the no battery charge warning light flashing red in his dashboard, he will show up at the repair facility and have his alternator exchanged. The car dealer will then turn in this and other old alternators to a remanufacturer together with other products, e. g. carburetors or worn out clutch disks, brake shoes and many more automotive units that can and will be remanufactured.

Returning for Remanufacturing

An automotive remanufacturer typically gets up to 80 % of his usable cores from his customers. There is a major problem for a remanufacturer mainly depending on exchange customers to supply his cores: Units exchanged in this way are almost always defective or broken in some way.

Defective Units

More cores come from specialized core brokers, who get their units from salvage yards or workshops where cars have been disassembled. These cores may even be more valuable:

That's because units supplied by core brokers were usually working when they were taken off the salvaged car and therefore require fewer spare parts or rebuilding efforts in the remanufacturing process.

Working Units

Professional Car Disassembly

Core brokers – or core suppliers – are professionals, who deal with salvage yards who disassemble cars as their key business.

Figure 27:
The Engine Units from
Retired Cars Are a
Valuable Core Source

Reliable Suppliers

Core brokers' main focus is to recover interesting units and parts and to have a core inventory on hand that enables them to supply the remanufacturing industry with the unit types, part numbers, and quantities they need.

Recovering all Remanufacturable Units

After having taken the complete engine unit out of the carbody, the core supplier takes apart further subassemblies and parts, such as gearbox, starter, generator, water pump, clutch, carburetor etc. to fill his stocks with sufficient numbers and types of units he expects to sell to a wide range of remanufacturers.

Figure 28:
Carburetors
in a
Core Suppliers
Stock

The leaders and the largest core suppliers even publish catalogs, out of which remanufacturers can order used units – whether one piece or a trailor load – just like ordering something from any other supplier.

Cores
from the
Catalog

Figure 29:
Cores and Part Numbers from the Catalog

Core Availability and Value

However, when a product, for example a specific car model, has newly entered the market, there are none or very few cores available in the beginning. There are very few (mostly accident) cars retiring in this early phase, and there are just the very first units becoming defective and being returned from the market. Cores are therefore rare and expensive in this first phase.

Core Prices

Later in the product's market cycle, as more and more cars are scrapped and the exchange business is running on a regular base, enough cores are in the pipeline and their price is becoming balanced. In the last phase until the end, when the car's market presence is declining, an excess of cores is occurring, with their prices rapidly falling.

Figure 30:
Core Availability and
Value Phases

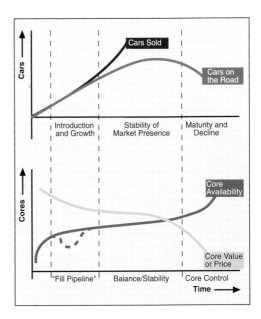

Of course there are several variables that influence the length and the characteristics of this core availability and value:

- Popularity of vehicle:
  Obviously if there are a lot of cars of the core's type out on the roads it will be easier to get remanufacturable parts.
- Years of original equipment (O.E.) production:
  This determines the length of that core cycle.
- Scrap rate of the vehicle:
  This factor is difficult to forecast. There are cars that come into the market and are scrapped very rapidly. There are also cars in the market that are today thirty years old and they are never scrapped unless they are totally demolished, depending on the style of car, etc.
- Vulnerability of the part:
  There are cars which are very hard on, let's say – a clutch. It wears out in a hurry. We also know cars which seem to be over-engineered and never need a new clutch. This type of clutch is never replaced, so that no core business is taking place at all for this particular model clutch.

Both on the core broker's and on the remanufacturer's side it is therefore important to be aware of one's actual position in the core availability and value cycle – especially towards the end of a product's market cycle in order to prevent oneself from accumulating too many cores that are surplus or from paying too much for the cores.

However, as our economies are going global, a special type of core which is available in sufficient or even excess numbers in one continent might still be rare in another part of the world.

Figures 31 and 32: Clutch Cores on Stock (above) and Turbo Charger Cores (below) Being Prepared for Shipment to the USA at a Large European Core Supplier

To balance core availability between various continents, the core supply business is also going global. Specialized core suppliers sell used European car units to the United States and vice versa.

Cores are moving all around the globe. Hard-to-get cores at one end of the world might well arrive from the other end, being the very welcome valuable and indispensable starting point of the remanufacturing process itself.

Figures 33 and 34: Rack and Pinion Steering Cores (above) and Car Electric Control Unit Cores (below) at the Same Site

Remanufacturing's Five Key Steps: Valuable Know-how plus New Technologies

Series Remanufacturing

The remanufacturing process itself takes place in a factory environment. It is organised as an industrial process, so that it can benefit from the advantages of a series production. These are a constant quality level and rationalization. Remanufacturing is carried out in the following five steps:

Figure 35:
The Five
Key Steps of
Remanufacturing

During each step, especially during reconditioning and reassembly, adequate quality assurance measures are also applied.

Quality Assurance

In particular the final testing of the product is sometimes claimed as an independent procedure which might deserve the status of a "sixth step". However, following the philosophy of understanding remanufacturing as a process equal to manufacturing, where standards define all quality assurance measures as an integrated part of the manufacturing operation in concern, one should consequently not isolate final testing, but regard it as an essential part of the fifth step that assures the quality of the product.

Equal to New

In any case, remanufactured products arrive at their customers with the same quality level, performance, endurance and warranty like a new product.

Valuable Experiences

As remanufacturing applies many principles of manufacturing, it also benefits from valuable manufacturing know-how. This not only applies to quality assurance. Experiences with production technologies and machine tools, or with assembly sequences and equipment known from manufacturing processes can also be transferred to remanufacturing purposes. This is true mainly for the two last steps which are component parts reconditioning and product reassembly.

New Solutions and Standards

The first steps of the remanufacturing process, namely disassembly and cleaning, are new technologies on an industrial level. Here remanufacturing itself has set new standards and plays the pioneering role of creating new solutions and adding technological know-how towards the closed loop from an old to a like new product.

Technological Summaries

In the following each of the five manufacturing steps is briefly summarized, dealing with the main challenges, technologies and an outlook towards future developments. These explanations mainly address the technologically interested reader and are given in a common sense for various automotive and electronic products.

**General
Outlines**

After that, facts and figures outlining the scope and scale of remanufacturing operations divided into main product lines, but nevertheless of a more general interest, will then be presented in two more chapters with many additional insights into the factories.

Disassembly

**Down
to the
Single
Part
Level**

As a prerequisite of all further steps to remanufacture a unit to its like new condition, in this first step it is completely disassembled to the single part level.

Of course this applies only to such an extent where joinings can be loosened without damaging or destroying a part. Epoxy seals of electric windings, spot welds, joinings obtained by pressing, forming, forging etc. cannot be disassembled.

Fortunately, many mechanical designs show a majority of joinings that can be loosened non destructively. The analysis of the 534 single operations which are necessary for the complete disassembly of an automotive engine showed the following result:

There were only 8 joinings (less than 2 %) obtained by pressing or riveting, the majority were easily unscrewable parts or just disassembly tasks like taking apart two components, e. g. pulling a piston out of its boring.

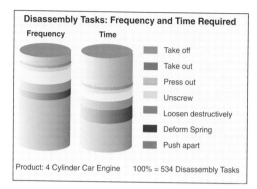

Disassembly Tasks: Frequency and Time Required

Frequency | Time

- Take off
- Take out
- Press out
- Unscrew
- Loosen destructively
- Deform Spring
- Push apart

Product: 4 Cylinder Car Engine   100% = 534 Disassembly Tasks

Figure 36:
Analysis of 534
Operations for the
Disassembly of an
Engine

For the rationalization of disassembly tasks, power tools and other mechanization devices are widely used in practice.

Mechanization of Disassembly

Figure 37:
Power Tool Assisted
Disassembly of a
Clutch

Innovative remanufacturers are developing their own new technological solutions for disassembly. Specially designed disassembly equipment does not always mean to handle the power tool, but can also result in passing the unit to be disassembled itself along specially engineered mechanized stations.

Figure 38:
Mechanized
Disassembly of
Automotive Units
– Principle –

Figure 39:
Mechanized
Disassembly of
Automotive Units
– Realization –

More than
Reverse
Assembly

Disassembly means much more than just reverse assembly. This is a fact, because there is no easy reverse operation for assembly operations like gluing, riveting, pressing, welding.

The disassembly task also includes an identification and immediate scrapping of parts, which are apparently not reconditionable like broken housings, burnt windings etc. It also includes the separation of all components which are fundamentally not reusable like gaskets, rivets etc.

Disassembly is also more difficult than assembly because dirt, rust and oil can cause the job of the workers to be slowed down. This encourages efforts to develop new solutions for the mechanization or even automation of disassembly operations. In recent years, also some experiments with industrial robots for disassembly operations have taken place.

Figure 40:
Telephone
Disassembly by an
Industrial Robot

Results from such experiments showed, that disassembly by an industrial robot might well be technically feasible for certain suitable products with many screws like telephones, engines or gearboxes. However, such approaches always quickly reach their economic limits and constraints due to the necessary large batch sizes for robot operation. Those are not available in most remanufacturing operations and would therefore lead to costly efforts and downtimes for continuous resetting of the equipment, tools, jigs and fixtures etc. Technical downtimes due to damaged or non original screws in the product etc. will also occur. In practice therefore, manual or

Robots for Disassembly

moderately mechanized disassembly will remain the adequate solution for the future.

**Further
Progress**

Further progress and some relief from todays' difficulties in disassembly can be expected however from the efforts noticeable towards product design for disassembly, which will be discussed in the relevant chapter later.

Cleaning

**Cleaning
and
More**

The second step in the remanufacturing process is the cleaning of all parts coming from the disassembly process to their reconditionable or reusable condition.

Cleaning is much more than washing away dirt and dust from the parts. It is also de-greasing, de-oiling, de-rusting and freeing the parts from old paint.

For this variety of purposes and complexity of the task many cleaning methods have to be applied subsequently or simultaneously.

These methods include washing in cleaning petrol, hot water jet or steam cleaning, chemical detergent spraying or chemical purifying baths, ultrasonic cleaning chambers, sand blasting, steel brushing, baking ovens and many more.

Remanufacturers' cleaning processes are becoming more "environmentally friendly" as they move to newer and more efficient cleaning technologies. They do not want to generate hazardous wastes because it is not good for the environment and the cost of disposal is skyrocketing.

Figure 41:
Clutch Discs Before
and After cleaning

Many cleaning processes, comparable to disassembly, can not be derived from common manufacturing processes and have required the development of new technological solutions by remanufacturers and their equipment suppliers.

Innovative Cleaning

Cleaning technologies also underwent significant changes in recent years, successfully moving away from chemical detergents with environmental impacts such as the ozonelayer-depleting chlorofluorocarbons (CFCs). Many remanufacturing companies are pioneers who have to be honored for having shown, that the replacement of CFCs which was for a long time considered not to be replaceable by other cleaning processes, is feasible. Additionally it can lead to similar or even superior cleaning results at lower cost, if the know-how to run other cleaning equipment and processes is developed.

Significant Changes

CFC-Free Cleaning

Among these developments one can also observe the revival of cleaning processes with water soluble detergents, which some years ago seemed to be phased out by chemical detergents like CFCs. Now the CFCs themselves are in fact phased out, while hot water and steam cleaning is on the upswing again.

Figure 42:
Hot Water
and Steam
Cleaning
of Engine
Parts

Trends
and
Parameters

Forecasting a more general trend in cleaning technologies, it is worthwile to have a look at the share of the four parameters which altogether contribute to the cleaning result.

The four cleaning process parameters are:

1. Chemical Effects (e.g. Detergents)
2. Temperature Influence (e.g. Heat)
3. Mechanical Action (e.g. Brushing Water Jet)
4. Time (e.g. Duration of Process).

There are clear indications that chemicals will be among the losers and mechanical action, including ultrasonic vibration will be among the winners of future cleaning process parameters.

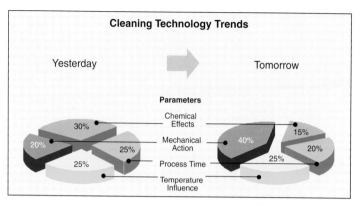

Figure 43:
Future Trends in
Cleaning

Further good news come from experiences with mechanical cleaning by glass bead or steel shot blasting.

Figure 44:
Steel Bead Blasting of
Parts in a Rotary Table
Cleaning Machine

Glass bead or steel shot blasting not only brings back a shiny surface of the part again, but also hardens the surface. This results in a better resistance against abrasion of the remanufactured product's parts than of the original new parts.

**Improving
Part
Quality**

Inspection and Sorting

A further step of great importance in remanufacturing is to assess the condition of the disassembled and cleaned parts as to their reusability or reconditionability. This task has two aspects:

Objective Criteria

First, the definition of objective criteria and condition characteristics to determine the condition of the components. Second, the development and application of suitable and affordable testing equipment. As the "like new" criteria, or the respective condition testing equipment from new manufacturing will only apply later after the reconditioning processes, at this point part assessment mainly relies on visual inspection.

Visual Inspection

For this, however, magnifying glasses, microscopes, high resolution cameras with electronic image processing and other devices already exist from comparable tasks elsewhere. They are of great help also for inspection tasks in remanufacturing.

Figure 45:
An Electronic Chip in Front of the Camera is Assessed on the TV Monitor Regarding the Soldering Joints

Besides optical appearance, many part condition characteristics can be inspected using objective technological criteria. Spring tensions can be measured, hidden leakages of housings etc. can be detected by air pressure or underwater tests, electric shortcuts can be measured etc.

Measuring and Detecting

Figure 46:
Measuring the Spring Tension of Clutch Plates

Figure 47:
Three-Dimensional Measuring of a Constant Velocity Drive Part

Figure 48:
Inspection
of Brake
Calipers

Sorting Parts

Depending on the various inspection results, parts are being sorted into three classes

- reusable without reconditioning
- reconditionable
- not reusable or reconditionable.

Besides this sorting process, considerable effort is often required for the manual sorting of similar, but not identical components, e.g. screws.

A facilitation of these efforts can on one hand be expected from using gauges for screw diameters, lengths etc. instead of visual separation of different parts.

Significant help can on the other hand also come from standardization efforts in the manufacturing industry, aiming at a reduction of part numbers and variants for even more different final product applications.

Reconditioning

Reconditioning is the remanufacturing step ensuring a like new condition on the part level again. It is the most important step in many applications. Depending on the product or unit remanufactured, it can occupy up to one half of the workplaces of a remanufacturing plant, as a case study of automotive engine remanufacturing has shown.

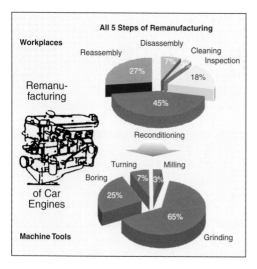

Figure 49:
Workplaces in Engine
Remanufacturing

For the remanufacturing of in most cases only slightly worn components, metal cutting processes such as turning, milling and grinding are preferably used.

Figure 50:
Valve Seat
Remanufacturing

Matching
Original
Tolerances

The equipment for this task, like lathes, milling and drilling machine tools, machining centers, grinders etc. is more or less the same as in manufacturing plants. As batch sizes of remanufacturing are smaller, the amount of manual labor is higher and the degree of automation is lower in remanufacturing plants. There are also cases however, where the original transfer lines used for the former new parts production have been moved to the remanufacturing plant and are now working for parts remanufacturing.

Through metal cutting reconditioning processes the geometrical dimensions of the parts will change, e.g. the diameter of a crankshaft through the grinding process. The remanufactured engine will receive bearings of slightly larger diameter to match the original tolerances. In many cases such geometrical changes stay within the original tolerances and have no influence at all. The collectors of starter and alternator rotors are perfectly smooth again after the turning reconditioning process. They not only look glossy as new, but will of course deliver the same powerful

Figure 51:
Rotors of Automotive
Electrics after
Reconditioning

current to the carbon brushes as they did in the unit`s first life cycle, although they have become slightly smaller in diameter.

Besides geometrical reconditioning, materials` properties are also restored to original standards by processes like hardening for example which are also known from new parts production. After professional reconditioning one can not even tell whether it`s a new or a remanufactured part.

Figure 52:
Remanufactured or
New?

If high optical demands are put on products, surface treatments such as chrome galvanizing, spray painting, powder coating etc. are also carried out as important remanufacturing processes.

Surface Treatments

Figure 53:
Freshly Coated Engine
Parts and Cowlings
Arriving for
Reassembly

New Spare Parts

Reusable parts obtained by the many different reconditioning processes represent a considerable portion of the components in remanufactured units and products. Non-reusable parts have to be replaced by new spare parts. To save new parts, the number of disassembled units can be increased in relation to the number of reassembled units, in order to receive enough reusable parts, at least the most important ones, from cores instead of more expensive new parts sources.

Reassembly

Assembly Lines

The reassembly of the parts to remanufactured products takes place on small batch assembly lines, using the same power tools and assembly equipment that used in new product assembly operations.

Figure 54:
Final Assembly of
Alternators

The assembly procedure is followed by a functional inspection or test run of each remanufactured product.

Test Run

Figure 55:
Constand Velocity
Drive Test

By this 100% inspection the reliabiltity of remanufactured products, which can be observed in later operation, is often higher than of similar new products inspected only by random sampling, and at any time higher than the reliability of just "repaired" products.

Remanufacturing Cost

Full Technology at Half Price

Since remanufacturing starts from the core (used part), not from ore and new parts like manufacturing, remanufacturing operates and turns out its products at lower cost than manufacturing. Of course individual figures vary significantly, depending on the industry sector and products remanufactured. One can consider remanufacturing to realize the same product at about half of the cost of a new one. Overall, remanufactured products are sold in a price range between 40% and 80% of a new product price, with an average of 60%. This complies not only with the cost but it is a win-win situation for both the customer receiving an attractively priced product and the remanufacturer being able to run his business profitably.

Figure 56: Cost Distribution between the five Key Steps

Looking at relative cost, there is a cost distribution between the five key steps of remanufacturing that also receives its main influence, naturally, from the type of product or unit remanufactured.

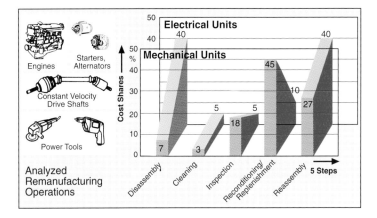

Generally speaking, it can be stated, that complex mechanical units, e.g. automotive transmissions and engines, show a main cost share at the parts reconditioning and new parts replenishment step. Electrical units, where only limited parts reconditioning takes place, show the disassembly and assembly steps as main cost factors. Assembly is generally an expensive task of remanufacturing, as it is in new manufacturing.

Remanufacturing at Work: Visitors Welcome

Two Thirds of all Remanufacturing

Remanufacturing has its strongest tradition and currently also its strongest representation in the automotive sector. Automotive products remanufacturing accounts for two thirds of all remanufacturing. In the USA remanufacturing is a fifty three billion dollar industry. Troughout the world, it can be estimated a hundred billion dollar industry at least.

Close to 100%

While in the European countries the remanufacturing business is growing very fast, in the USA it has already reached as much as 45 to 55 percent market share of the replacement parts in the automotive aftermarket. Depending on the part or unit, this figure might well rise close to 100%. For example, replacements of starters and alternators reach a remanufactured share of 90%, no matter whether the car with the exchanged unit is being serviced in America, in Europe or in Asia.

Customers Worldwide

It`s certainly no question of different national cultures, that car owners all around the world prefer to purchase a remanufactured unit for 35 to 50 percent below the price of a new one. If a car has broken down and can be put back into service with an exchange unit on the same day, at lower cost but with the same warranty as new, any customer will prefer a remanufactured unit.

Complete Alphabet

Listing just the most important of the various kinds of automotive units being remanufactred, one finds the whole range of units which stand for function, performance, safety and comfort in today's cars and trucks: In alphabetical order, among these there are

- Air Brakes
- Air Conditioners
- Alternators and Parts
- Anti-Lock Brake Systems
- Brake Shoes
- Brake Cylinders
- Calipers
- Carburetors
- Clutches
- Constant Velocity Drive Shafts
- Cruise Controls
- Cylinder Heads
- Differentials
- Electronic Control Modules
- Engines and Parts
- Fan Clutches
- Front Wheel Drive Axles
- Fuel Injectors
- Generators and Parts
- Master Cylinders
- Oil Pumps
- Power Brake Units
- Power Steering Gears
- Power Steering Pumps
- Power Window Motors
- Rack and Pinion
- Smog Pumps
- Starters
- Steering Units
- Torque Converters
- Turbo Chargers
- Transmissions and Parts
- Water Pumps
- Wiper Motors

Although just about everybody is a regular customer of the remanufacturing industry, surprisingly almost nobody has ever had a look into a remanufacturing plant, whether in real life or through one of the rare television or newspaper reports about remanufacturing. It`s worth having a look, however. Most remanufacturers like to open their doors and to welcome visitors, just as many manufacturers do. The following insights offer some impressions from various automotive remanufacturing operations.

A Look into the Plants

Engines

One out of ten cars and trucks on our roads, on an average , needs a replacement engine during its more than ten year life. The automotive after-

Ten Percent Remanufactured Engines

Upgrading Efficiency

market industry itself as well as thousands of independent remanufacturers rework worn out or defective engines back to Original Equipment performance standards or even beyond: Advances in technology which occured in the meantime like better fuel efficiency, lower emissions etc., might be incorporated in the engines which return from a first service.

From Light Weight to Heavy Duty

Engine remanufacturing reaches from exclusive sports cars to everybody`s mass compact vehicle, from gasoline to Diesel operation , from light weight roadsters to heavy duty trucks.

Figure 57:
Engine Disassembly

Figure 58:
A Like New Remanufactured Engine

In particular in the heavy duty sector, remanufactured engines prove both their reliability and durability as well as their flexibility:

They have often run for hundreds of thousands of miles during their first service period. But they are ready for the next half million miles after a professional remanufacturing process. Regarding flexibility, an engine returning from its first life cycle in a truck, might well do its next duty in an excavator, compressor or power generator. Or it can move out from the oceans, where it has powered a ship for many years, onto the railroads as a locomotive engine after remanufacturing.

Figure 59: Remanufactured Engine for Railroad Service

Engine remanufacturing, which reaches around five million units anually, might not belong to the remanufacturing operations highest in numbers. Considering value however it certainly counts as one of the most important contributions to keep the large fleet of half a billion cars and trucks on our planet's roads.

Starters and Alternators

Any car or truck relies on two indispensable electric units: Starters and alternators.

It does not need those two only – on an average, it also needs two of each during its life cycle. One day, the car won't start in the morning. The other day, the driver will suddenly notice a red light on his dashbord during a trip. Both means the starting point for remanufacturing.

Twice in a Lifetime

Thousands of companies remanufacture starters and alternators as their basic business. Every core supplier has them in their product line.

Figure 60:
Starters and
Alternators: Ready for
Remanufacturing

In many cases, especially with alternators, there is just a small electrical defect, such as a broken contact in the diode plate. Nevertheless, the unit will receive a full both electrical and mechanical overhaul, e.g. receiving new bearings, carbon brushes etc. during remanufacturing. Thus, a remanufactured starter or alternator has the same life expectancy as a new one. The only difference is its price. Customers feel extraordinarily

Figure 61:
Remanufactured
Alternators: Full
Technology at Half
Price

happy about paying 50% less here, as electric breakdowns come suddenly and unexpectedly. They are not foreseeable like with brakes or clutches for example.

Regarding numbers, starters and generators are the shooting stars of remanufacturing. They approximately equal car manufacturing numbers, which are 50 million units per year worldwide. Only some kinds of brake shoes can reach such high numbers, as they wear out several times during a car`s lifetime.

Clutches

Clutches belong to the parts with the toughest jobs in a car`s or truck`s everyday life. Stop and go traffic, pulling heavy trailers, starting from a hillside red traffic light. Any driver knows such situations he might hate or not – but certainly his clutch does. Earlier or later, it will wear out under such severe conditions.

Tough Jobs

Remanufacturing of clutches is the answer to this problem as well. It is a profitable business both for big OEM`s as well as for many small and medium sized remanufacturers.

OEMs and SMEs

Figure 62:
Clutches on the Move
– Ready for
Remanufacturing

**Proven
Equipment**

Clutch remanufacturing is a tough job too, when it comes to loosening rivets, managing powerful spring tensions, and handling the heavy parts of trucks.

*Figures 63 and 64:
Clutch Disassembly
and Assembly*

Fortunately however, there are cranes, machine tools like presses and many other factory equipment that help with that.

Clutches are completely disassembled and cleaned, their discs are stripped from old linings and receive new ones. Pressure plates receive a new grinding to a mirror-like shape again. All springs are measured regarding length and tension or are replaced where necessary.

Function Test

Carefully reassembled and undergone a comprehensive function test, a remanufactured clutch is ready for its next tough job. Clutch remanufacturing is also a business of fast movers. Remanufacturing numbers count for several millions annually worldwide.

Electronic Control Units

Electronics mark the most exciting progress in automotive technology today. Electronic control units, or computers, have conquered nearly every important function of the modern car, and they will continue to do so.

Electronic units control engine and drive train performance, such as electronic fuel injection, transmission control, anti-slip functions and more.

Engine Control

Computers are used for safety features, such as anti-block brake systems, airbag ignition in case of an accident, distance alert to prevent an accident etc..

Safety Control

Electronics have also become indispensable for driving comfort and many functions nobody thought about just a decade ago. Air conditioning and audio entertainment have been known for a while as some luxury inside the passenger cabin. Now mobile electronic communication and car navigation systems offer further attributes and assistance from the outside world. And this is just the beginning of many more electronic functions.

Comfort Control

Visionists see a future car knowing its way and driving there through heavy traffic electronically controlled by itself one day.

The Electronic Car

Staying on ground however, this still means when there is a new function , there can also be a failure. So remanufacturing of electronics is on the upswing.

Figure 65:
Electronic Control
Unit Remanu-
facturing Centre

Most car manufacturers or their Original Equipment electronic suppliers are also busy electronic control unit remanufacturers.

Figure 66:
Electronic Control
Unit Diagnostic
Test

Figure 67 (below):
Electronic "Freezing"
Test at 30 Degrees
Below Zero

And independent market players are coming up, too. The necessary knowhow and the equipment for electronic control unit remanufacturing is in most cases available from new manufacturing. But some tasks, such as diagnostics and failure detection / analysis are new.

For this, dedicated new hardware and software also have to be developed. This is a chance for new specialists and offers innovative business opportunities for any participant in the new remanufacturing cycle closing the loop also for electronic units.

Figure 68:
Testing a Car
Computer with a
Personal Computer

Figure 69:
Temperature Test
Chamber from
– 30° to + 80° C

Rapid Innovation   As product innovation and new technology development of electronics proceed at the fastest speed throughout all industry sectors, the same car model might well be equipped with two or more subsequent generations of different electronic control units for the same function during its production period. In the aftermarket this also means an explosion of the variety of electronic control units to be serviced and remanufactured. There are OEM`s, who already handle more than one thousand different electronic units within their remanufacturing operations.

Record Breakers   Thus, this field of remanufacturing activities will certainly never be a record breaker in numbers per unit, but remanufactured electronic control units are ready to set an all-time high in product variety. A comparably wide range of different products can only be discovered elsewhere, if one dares glance beyond automotive parts into other industrial sectors of remanufacturing.

Remanufacturing beyond Automotive Parts: Discovering a Hidden Green Giant

In addition to the impressive activities in the automotive sector, technology development and market expanse in many more product and industry sectors are fruitful grounds for the idea of remanufacturing as well. This is not just an idea however. In practice one can in fact find a nearly unlimited range of products being remanufactured in a wide range of applications. The product spectrum reaches from remanufactured units with similarities to automotive parts, such as mechanical machinery and all kinds of electrical apparatus, via other mechatronic and electronic products, into many more independent and individual niches. They include office furniture, toner cartridges and tire retreading, to mention just a few significant ones. Lot sizes reach from batch size one up to hundreds of thousands of units. Product values show dimensions from a one dollar electronic component up to a hundred million dollar Jumbo Jet undergoing the remanufacturing-alike "D-check", the most comprehensive overhaul.

Fruitful Grounds

From One to Hundred Thousand

Machine Tools and Industrial Robots

Looking into some remanufacturing plants and starting with electrical/ mechanical machinery as a first application in the order mentioned at the beginning of this chapter, the remanufacturing of industrial robots is an interesting example.

Mechanical Products

Figure 70:
Industrial Robot at
Work

Ready for Remanufacturing

After years of heavy duty three shift work parbly in a forge shop or comparably severe conditions, mechanical wear as well as electrical / electronic behaviour might cause the industrial robot`s performance to become insufficient.

Remanufacturing the robot, reworking its moving mechanics and upgrading its control electronics is the best way out of this problem.

Figure 71:
Robot Arms during
Remanufacturing

If one has a look into the robot remanufacturing activities in practice, they can be observed taking place at OEM`s.

Independent remanufacturing companies and the robot users, e.g. the automotive industry remanufacture robots as well.

Figure 72:
Robot after
Remanufacturing

Remanufactured robots, just as a representative for a wide range of other remanufactured factory equipment like machine tools, receive the latest generation of computerized control for their next life cycle. They are in fact better than new after the remanufacturing process – at a cost half compared to a New One.

Latest Equipment

Vending Machines

An interesting example of mechatronic products, being widely used and remanufactured, is the field of vending machines for hot or cold drinks, candies etc. These products, e.g. vending machines for cold drinks, contain a wide variety of components and parts which are also used in automobiles and household appliances, such as a housing nearly as big as a car, an electronic money changer control, a refrigeration and cooling system, an electric drive and control system.

Mechatronic
Products

The products are disassembled, reconditioned and upcycled to the current state of the art.

Figure 73:
Vending Machines
before
Remanufacturing

Figure 74:
Vending Machines
after Remanufacturing

Upcycling

These products are upgraded or "upcycled" as well. Their appearance meets the current designs and their technical performance, e.g. money changer, often offer new functions which have not been included during the product`s first life cycle.

Copying Machines

Copying machines of various brands, are another important example. They are rebuilt and upcycled either by independent remanufacturers or OEM`s.

One of the big OE remanufacturers in Europe runs a remanufacturing center that successfully turned out more than 60000 copiers annually in the mid 90s. They call this activity not just a "remanufacturing center" but a business unit "ARO" which means "asset recovery organization". This demonstrates an ecological attitude and happiness not to spoil valuable natural resources.

Asset Recovery

Figure 75: Remanufacturing of Copying Machines

If OEM`s remanufacture the product themselves they often invest into – and profit from – modular and intelligent product designs to ease future upcycling. Pioneers in this field justify these investments and design efforts – for example in the direction of an improved product disassemblability – by explaining:" We won`t improve our product design to sell the product once and then tear it down more quickly – but we do

Multi-Life-Designs

improve its design, because we sell the product twice – once new, later rebuilt." And, in addition, "A product, which can easily be disassembled, can also be assembled more easily." This issue will also be discussed in a later chapter.

Electronic Products and Parts

Quantity and Quality

If one has looked into the product return collection centers and used products` warehouses in the electronic industry, one is impressed both by quantity and quality.

Regarding quantity, as Personal Computer production volumes have now overtaken cars in numbers, it is a surprise to realize how fast such locations fill up with look like new used electronic products.

Figure 76:
Used Computer
Monitors Warehouse

Still Working

Used electronic products do not just look like new. In fact, 80% of them are still working. They have been overtaken technically by innovative new products. As they are still working, this offers an ideal base for remanufacturing

including their modernization by upgrading/
upcycling.

Personal computers and other products equipped
with a keyboard, such as supermarket cashier
equipment, are remanufactured widely and frequently.

Figure 77:
Supermarket Cashier
Equipment during
Remanufacturing

One of the most recent examples of remanufacturing in global scope and scale are cellular phones, which are remanufactured at a European
Center of a Japanese OEM.

Figure 78:
Cellular Phones – the
Most Recent Addition
to the
Remanufacturing
Business

If the whole electronic product cannot be remanufactured as such, reuse, remanufacturing and good-as-new remanufacturing of electronic components is also an important solution approach.

Traditional Unsoldering of Electronic Components

Unsoldering of valuable electronic components, such as processors and memory chips from electronic circuit boards has been known for a while. This process however, which has traditionally been carried out by putting the board into a hot lead bath to loosen the components carefully from them, is being critized because of the possible temperature shock for the components.

Innovative unsoldering technology therefore uses soft infrared light warming instead of a lead bath.

Figure 79 (above): A Worker Standing at a Lead Bath

Figure 80 (right): Infrared Unsoldering of Electronic Components for Reuse

This technology has no risk at all for the components to be reused – and there are many of them. The reusable component spectrum reaches from a half dollar capacitor via PC upgrade kits up to a two hundred dollar phone microprocessor board. The market leaders in electronic component remanufacturing or retrade are small and medium sized companies with around 100 employees, handling 10,.000 different component numbers and have one million components in stock.

Figures 81 and 82: Electronic Component Remanufacturing and Retrade is on the Upswing

Office Furniture

Office Furniture, a product with a strong tradition and also millions of units installed in our offices, belongs to the most recent discoveries of the remanufacturing industry and very profitable as well: Office furniture remanufacturing shows one of the fastest growth rates of all remanufacturing businesses. In the USA it is growing about 25% a year.

A Recent Discovery with Fastest Growth Rates

Figure 83: Remanufactured Office Furniture is Good as New

Figure 84:
The Long Life Basic
Structure of Partition
Panels Receives New
Fabric

Office remanufacturing activities cover all the products that can be found in our companies' offices and work station cubicles: Desks, desk chairs, filing cabinets, partition panels, meeting tables and chairs.

The remanufacturing process varies for each type of office furniture. Desktops receive new surfaces. The old ones are shaven off or, if in good condition, just sanded down. Then a new plastic laminate is applied to the work surface. Metal parts receive a new spray paint or powder coating surface. Filing cabinets are remanufactured like a car body's parts after an accident. Remanufacturers hammer out dents, replace handles and screws, ensure that all moveable parts function properly and repaint or powder coat the unit. Partition panels receive new fabric covers. Metal frames are inspected and visible metal parts receive a fresh coat of paint – and often they are completely recovered to better fit with current styles and colors. Chairs receive new fabrics and also new cushions, if necessary.

Remanufactured office surfaces look identical and resist scratches as well as brand new office furniture, i.e. desks, cabinets and panels.

Figures 85 and 86:
Replacing the Fabric
of Chair Cushions

Both OEMs and many independent companies offer office furniture remanufacturing services. Customers can supply their own furniture for remanufacturing or simply purchase remanufactured furniture without any type of trade-in.

Figure 87 and 88: Work Surfaces before and after Remanufacturing

Customers of remanufactured office furniture realize savings of up to 50% of the cost of purchasing new office furniture. It is a real win-win situation for all participants and for the environment as well. Therefore, within only ten years, remanufactured furniture has grown from a small segment of the office furniture retailing industry to a 10% share of the overall nine billion dollar commercial furniture business in the USA. Some experts in the industry predict that remanufacturing will command a 25% market share within just another five years. Office furniture might well become the new shooting star of remanufacturing activities beyond the automotive industry.

25% Market Share Predicted

Hidden Giant

Huge and Hidden

Altogether, the remanufacturing industry, if seen as an aggregation of companies producing a broad array of like new products from old products, is both huge and hidden. With its wide range of remanufacturing activities, this industry makes major contributions to the economy, employing hundreds of thousands of people, creating billions of dollars of products, and paying taxes on the profits.

In fact, the remanufacturing industry, due to its many product sectors has not presented a united presence to the public so far, is a "Hidden Giant".

Recognizing Remanufacturing

This is also the title of a recent study under the direction of Professor Robert T. Lund from Boston University, which for the first time recognizes the remanufacturing industry in the United States for its economic importance.

Remanufacturing counts for

- more than 70,000 companies
- over 50 major product categories
- about 53 billion dollars annual sales
- about 500.000 direct employees
- about 25 people average company size.

Worldwide, if one derives and extrapolates the very valuable data collected in Prof. Lund`s database with 9,903 remanufacturers surveyed, the above figures would increase by 50% to 100%. Remanufacturing creates hundreds of thousands of jobs. Remanufacturing is also a Hidden Green Giant. It has to be appreciated for its contributions to the environment and to societies in terms of energy, materials, labor and capital equipment conservation, and reduction of waste.

Remanufacturing Professionals: A Job Machine Accelerates

Labor Intensity

Remanufacturing is a labor-intensive business. In most cases, it requires three to five times the people required to manufacture the same product. What are the reasons?

First, disassembly, cleaning, inspection and sorting of parts are additional work in the remanufacturing process. Plus, not to forget, there are careful and thorough qualitiy assurance measures throughout all steps.

Second, and maybe more important, renaufacturing batch sizes are much smaller than in new remanufacturing. The degree of automation is adequate: There are workers instead of robots, assembled units are transferred by hand instead of conveyors etc.

Figure 89: Remanufacturing Workers Like Their Job

Good News

Keep this in mind: Even though it is a labor-intensive business the good news is that remanufactured products still arrive 50% cheaper than new ones. They don`t need their expensive parts to be produced, as those come from old units. There is also good news regarding the labor market. Remanufacturing creates jobs – it even has the ability of capturing jobs that have been lost to cheap manufacturers in developing countries.

Skills and Abilities

Qualified personnel are the greatest assets of the remanufacturing business. Reassembly and quality assurance workers and professionals need to have the same skills and abilities as in manufacturing – or even beyond. Parts`conditions vary to a higher extent in remanufacturing than in manufacturing – so professionals need a sure eye to decide about them and to manage them adequately.

Also disassembley personnel, although sometimes considered to do just a dirty job, determine the units remanufacturability in the very beginning. They have to work carefully. Otherwise the product might well have survived its former owners mistreatment, but could be damaged during disassembly and handling.

Figure 90:
Careful Application of
a New Surface

But remanufacturing workers like their job – they are motivated by their contribution to a healthier environment by performing the ultimate form of recycling. They know about the attractively priced, perfectly customer oriented service their work turns out.

Watching a worker in the office furniture remanufacturing industry, one world rather consider him caressing his part than just applying a plastic laminate sheet to a reprocessed work surface core.

Many so far unskilled workers like remanufacturing also for another reason. It has changed their lives. For some of them it has even meant moving from welfare to work. They started with easy tasks, grew their skills and are now proud to have become a reliable expert for a certain remanufacturing operation within their company. Besides blue collar work, there are also specialized mechanical and electrical engineers in remanufacturing design, research and development departments. And there are additional retail jobs for remanufacturing operations.

New Jobs

Remanufacturing is a job machine, and it is accelerating. Europe and Asia each count for tens of thousands of remanufacturing professionals. The US Remanufacturing Industry Giant shows a direct employment of about five hundred thousand people. With or without adding indirect employment, economies and societies have become millionaires and winners from remanufacturing.

Product Design for Remanufacturing: The Secret of Extra Benefits

The product design determines two thirds of the remanufacturability of a product. Intelligent designers can therefore considerably inprove the suitability of a product or unit for disassembly, cleaning and inspection as the main new issues. Design for manufacturability and design for assembly have already been known for a while. Generally speaking, design for recycling and remanufacturing are key elements of an overall concept of product design for the sustainability of our environment.

Figure 91:
Design for
Sustainability Also
Means
Remanufacturability

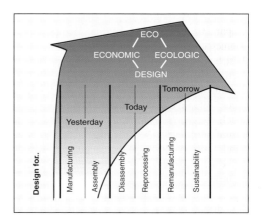

Harmonizing Economy and Ecology

In this surrounding, "Eco-design" has also become a frequently used slogan, when it comes to design and engineering activities regarding a better protection of our environment, when the design of products and processes is being considered.

With the current green movements, it is no surprise, that nearly everybody immediately interpretes the abbreviation "eco" as primarily "ecologic", – however, one should not forget, that it can stand equally for "economic" – and, in fact, should do so. Fortunately, there is an increasing range and number of examples from automotive, electric and electronic product designs, where "eco" really represents both:

Innovative eco-design does not solve challenges with combined ecologic and economic tasks by finding a more or less satisfying compromise between conflicting aims, but by developing new solutions with progress and advantages on both sides.

Advantages on Both Sides

Apparently, design efforts to improve the "environmental behaviour" of a product during the first two life-cycle phases which are the manufacturing and the use phases, have already seen considerable progress both traditionally and recently. On the one hand, avoiding waste or too much energy consumption saves cost at the manufacturer's or user's expenses. On the other hand, because there have been quite a few legislative acts recently e.g. to reduce emissions from factories, cars, etc.

Eco design to improve the products' "environmental behaviour" during the third phase, the recycling including remanufacturing, has gained interest and importance mainly in the past few years. The designer can improve the suitability of a product for recycling (but also its environmental features regarding the first two life-cycle-phases) by using the opportunities and

Eco Design for the Third Phase

making the right decisions in three dimensions (or design areas) of design for recycling: selection of materials; designing of the product structure / designing of joinings.

Figure 92:
Design for Recycling
and Remanufacturing

**Checklists
and Guidelines**

For each design area, standards, checklists and guidelines for recycling oriented design of technical products have been worked out in science and industry supported by the author. They are applied within the automotive, electric appliance and electronic industries throughout Europe, America and Asia.

**Design for
Disassembly**

One of these checklists, located within the third design area, the disassembly oriented design of joinings, clearly shows that it really supports the two-fold eco design approach. Design for disassembly and remanufacturing very often helps assembly and manufacturing too. To give an average figure, various experiences account for five percent reduction in assembly time for every ten percent progress in cutting disassembly time, e.g. by reducing numbers of screws, using snapfits instead of rivets etc.

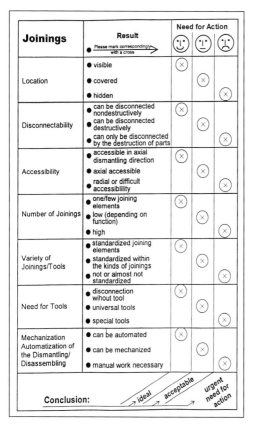

Figure 93:
Design Checklist for
Easy Disassembly

Joinings	Result ● Please mark correspondingly with a cross	Need for Action		
		☺	😐	☹
Location	● visible	⊗		
	● covered		⊗	
	● hidden			⊗
Disconnectability	● can be disconnected nondestructively	⊗		
	● can be disconnected destructively		⊗	
	● can only be disconnected by the destruction of parts			⊗
Accessibility	● accessible in axial dismantling direction	⊗		
	● axial accessible		⊗	
	● radial or difficult accessiblility			⊗
Number of Joinings	● one/few joining elements	⊗		
	● low (depending on function)		⊗	
	● high			⊗
Variety of Joinings/Tools	● standardized joining elements	⊗		
	● standardized within the kinds of joinings		⊗	
	● not or almost not standardized			⊗
Need for Tools	● disconnection wihout tool	⊗		
	● universal tools		⊗	
	● special tools			⊗
Mechanization Automatization of the Dismantling/ Disassembling	● can be automated	⊗		
	● can be mechanized		⊗	
	● manual work necessary			⊗
Conclusion:	→ ideal → acceptable → urgent need for action			

With more innovative design approaches, sometimes even revolutionary turnarounds in product suitability for manufacturing and remanufacturing occur. A stunning example comes from a new design concept for holding together the sub-assemblies of electronic products. The new design solution replaces the familiar metal and plastic housing structure by a foam housing for all the parts such as circuitboards etc. In this technology, all of the components are held by

A Turnaround by Eco-Design

their own geometry in form fitting spaces in the foam chassis. There are no screws, no rivets, no joining elements at all. After the cover is closed, everything is self-fitting.

Figure 94:
Innovative New
Design

If comparisons are drawn between the classic design and the new solution, the results are breathtaking. The new design concept allows

- a 70% reduction in housing mechanical parts
- a 95% reduction in screw joints
- a 50% reduction in assembly time
- a 90% reduction in disassembly time
- a 30% reduction in transport packaging
- a 50% reduction in time and expenditure for the mechanical development of the housing.

Eco-Design is a Major Innovation Driver

So, as this example impressively demonstrates, new design challenges, improvements and changes might just be caused and initiated by disassembly and remanufacturing issues, but their solution by breakthrough innovations can give interdiciplinary incentives and benefits to the designer`s work and – generally speaking – to all the industrial world.

Last but not Least: Quality and Safety First

Remanufacturers are responsible for the safety and reliability of their products – just as manufacturers are.

Product Responsibility

This is why quality control and quality assurance measures throughout the entire remanufacturing process with all its five steps is such an important issue. Remanufacturers take quality serious. Most of them therefore apply the same standards of quality and business management systems which are required and applied by original car manufacturers.

Many remanufacturers have achieved or are working toward the ISO 9000 Quality Standard and QS 9000 certification, which is an even higher standard of quality.

ISO Standard QS 9000 Certification

However, remanufacturing is still different from manufacturing. This is why remanufacturers have also worked out additional guidelines and self commitments, such as the Recommended Trade Practices of the European automotive parts remanufacturers. By undertaking such efforts and applying quality management systems and standards, remanufacturers are in the position to guarantee a thoroughly monitored and quality assured process and product.

Self Commitments

This is the reason why remanufactured products come with the same or even with a better warranty than new.

Warranty

Why are remanufactured products this reliable? What`s the reason behind some remanufacturers even granting life time warranties on their products? It`s not just because they have quality procedures during the process. There are also explanations from product maintenance and reliablility forecasting sciences and literature:

Discussing only the scientifically proven and most important fact, one has to look at the failure rates and product life time.

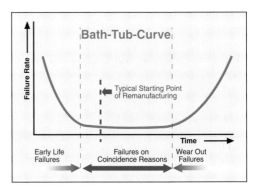

Three Failure Rate Phases

There are three phases of failure rates during a technical product's life:

First there are the early life failures – often at a surprisingly high rate in the beginning.

Early Life Failures

Everybody is familiar with the fact, that even a brand new car might not start already on its second morning. Due to material defects or a manufacturing error, an electrical contact in the starter has failed. The starter has to be turned in for remanufacturing.

Lowest Failure Rates

Then there is a second phase – a very long phase of lowest failure rates. The product or part is working for the duration it has carefully been designed for. Failures only occur on coincidence reasons, by accidents, extraordinary stress etc.

Wear-Out Failures

In the end, failure rates go up again – due to wear and fatigue of parts and materials. Due to its appearance, the graphic showing the whole three phase connection is called "bath tub curve" even in science.

The pleasing fact is, that old units arriving for remanufacturing still have their parts at the very beginning of the long and stable second phase of low failure and high reliability expectations. If their defective parts e.g. early failed electrical parts or worn out minor parts like carbon brushes or bearings are replaced, their main mechanical parts will easily withstand the stresses of the next use.

This is the main reason why remanufactured products are so reliable. Of course, one has to know a specific part`s position in the failure rate and life time connection to forecast its reliability for the next use period. Remanufacturers are aware of the theoretical knowhow and they have gathered practical experiences as well.

Knowing One's Position

Whenever they are not sure or consider the wear-out failure rate is close to happening, they will not reuse that part or remanufacture that unit.

If in Doubt: No

This especially applies to units and products relevant for a car`s safety for example, such as rack and pinion steerings, brakes etc.

Here the same principles have to be followed like with aircraft overhauling, which is a process carried out by any airline or aircraft mechanic and which is nothing else than remanufacturing regarding the most comprehensive so-called "D-checks" which include every step from disassembly to reassembly.

Aircraft Standards

Aside from theory, from the field of remanufactured products, there is also good news. They work to their customer's satisfaction.

Higher Rates

Remanufacturing's Success Factors: An Aid to Decision Making

Impressing Volumes

At present remanufacturing – the ultimate form of recycling – already exists in impressing volumes both in automative and non automative industries and markets. A forther growth in numbers and applications can be expected and should be encouraged. Where are the prerequisits, the success factors and the limits? Are there further remanufacturing business opportunities neglected so far?

As every product will retire at a time, the necessity of recycling is showing up in all industries, and a decision making is required.
The first question arising is entitled

- Recycling by Remanufacturing ("Upcycling")?

  or just

- Recycling by Reprocessing ("Downcycling")?

Conserving Values

Remanufacturing is the preferable process as it conserves the added value which has been put into the product by design and manufacturing. Reprocessing destroys all this value by shreddering and smelting it down and just recovering some natural resources.

Product Assessment

To assess a product's suitability for remanufacturing, all factors of technical, economical and ecological nature will influence the decision and have to be taken into consideration.

The suitability of products for remanufacturing can be examined by evaluating a range of eight different criteria:

- Technical Criteria (kind or variety of materials and parts, suitability for disassembly, cleaning, testing, reconditioning)
- Quantitative Criteria (amount of returning products, timely and regional availability...)
- Value Criteria (value added from material/ production / assembly)
- Time Criteria (maximum product life time, single-use cycle time...)
- Innovation Criteria (technical progress regarding new products vs. remanufactured products...)
- Disposal Criteria (efforts and cost of alternative processes to recycle the products and possible hazardous components...)
- Criteria Regarding Interference with New Manufacturing (competion or cooperation with OEM`s ...)
- Other Criteria (market behavior, liabilities, patents, intellectual property rights...)

In each field, an evaluation of the parameters in brackets has to be calculated for decision making between remanufacturing or reprocessing.

Portfolio Methods

Known from marketing strategy tasks and other fields, portfolio methods can be of some help and have been developed by the author in order to prepare decisions by developing scales for the evaluation and comparison of different criteria or of questions within certain criteria. As one example of such a portfolio, the evaluation of the time criteria mentioned above and the findings of suitable products, which in fact are already being remanufactured in practice, is carried out in a connection between maximum product life time and average product cycle time per

Figure 96:
Portfolio Evaluating
Time Criteria for the
Decision between
Remanufacturing and
Reprocessing

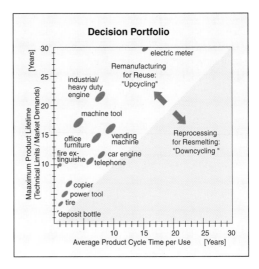

New Marketing Strategies

use.

There are clear indications, that the range of products or parts, as well as the possible markets, are still only at the beginning of the game called remanufacturing. This includes new marketing strategies such as leasing. Such an approach "Sell the performance, not the Product" is successful today in the field of copying machines.

An example, where this approach named "sell the performance, not the product" is successful already today, can be shown in the field of copying machines. The customers` needs can be defined as "to make 10.000 copies per month" (not "to own a copying machine"). Accordingly he does not buy the equipment, but has a leasing contract with the supplier, who guarantees that an up-todate copying machine with the required performance is always available in the customer`s office. From time to time this copying machine is exchanged for a "new" model – the latter having been remanufactured and upcycled in the OEM`s premises or by an independent subcontractor to him. In this field therefore new business opportunities arise also for small and medium sized regional enterprises.

While after sales service, leasing and rentals are known market fields for remanufactured products, the "ETN-Product" approach is still an exception. ETN means "Equivalent To New" and says that a product sold as a new product contains also remanufactured parts. This is a new business opportunity for small and medium sized

Equivalent to New

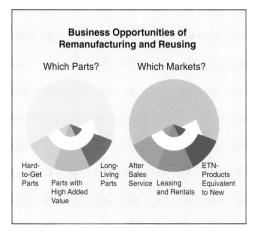

Figure 97:
Parts and Markets for
Remanufacturing

enterprises as well as for OEMs.

This is desirable and it will become common one day without making a great fuss about it. In the future, new products will contain remanufactured parts as they contain reprocessed material from the steel industry today.

Energy and Material Savings by Remanufacturing: A Star is Born

No Recycling of Energy

Energy cannot be recycled. It is being spent for whatever process is carried out to create a product. Then it is lost, or, to be honest, it is embodied in the product.

Much Energy for Material Recycling

As there are also materials embodied in a product, traditional recycling processes aim at conserving a majority of these materials by shreddering the product, separating its main different materials, smelting them down to recycled materials and producing new parts from them.

These recycling processes, besides giving up the energy that has been embodied in the first product, need a lot of additional energy to arrive at the next product.

More Energy for Manufacturing

At least however, their energy requirements are in most cases still somewhere below those of manufacturing new parts and products from all new material. Some material recycling processes nevertheless need even more energy to close the loop than processes which start from ore. There are also some emmissions that accompany many material recycling processes.

Much less Energy for Remanufacturing

Remanufacturing conserves the energy embodied in a product and, compared to all the above, nearly needs no additional energy. In fact, if one undertakes the effort of a careful and thorough comparison of energy requirements of remanufactured or manufactured products, this leads to remarkable results:

In a case study carried out by the author's research group, the most frequently remanufactured automotive units

- **starters** (5 types, heavy duty and passenger car)
- **alternators** (3 different types)

have been evaluated regarding the energy consumption of

- **manufacturing** or
- **remanufacturing.**

The same kind and type of unit was considered. In order to achieve fully realistic results, the comparison was based on the assumption, that both the parts for the newly manufactured unit as well as the amount of new spare parts running into the remanufactured unit are produced using materials with today's shares of recycled steel, aluminium, copper etc.

Realistic Assumptions and Results

The results were amazing: to manufacture one new starter on average requires more than eleven times the amount of energy of a remanufactured one. A similar situation can be found with alternators. Here a new one requires close to seven times of the energy to make than a remanufactured one. In other words these products can be realized with an effort of only 9% (for starters) or 14% (for alternators) of energy if remanufactured and not manufactured new.

Multiple Savings

Figure 98:
Regarding Energy and
Material
Consumption,
Remanufacturing is
Far Ahead of
Manufacturing

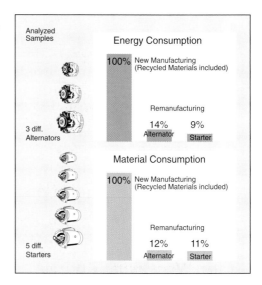

The results about material conservation by remanufacturing were impressive as well: One new starter on average requires more than nine times the amount of new material needed than a remanufactured one. Or, new alternators require more than eight times the amount of raw material than is needed for remanufacturing the same units from old ones.

So, with the same amount of energy or material given, one could make from 7 to 11 units more by remanufacturing them than by manufacturing.

An Extrapolation to World Scale

Figure 99 (right page):
Energy Savings by
Remanufacturing
Worldwide

All over the automotive industry and all across many other industries, many more units and products are being remanufactured. Using statistical data and calculating the scientific data about materials and energy intensity of processes used for manufacturing or remanufacturing all those products and units, one can extrapolate the overall energy savings by remanufacturing.

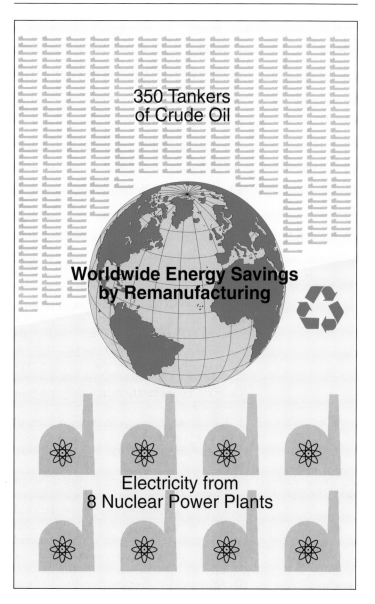

Worldwide Material Savings by Remanufacturing: A Fully Loaded Train 3.000 km/1.600 miles Long

As a result of such an extrapolation it can be stated:

– **Energy:** The savings of energy by remanufacturing amounts worldwide to 120.000.000.000.000 BTU`s (British Thermal Units) a year. This is an amount of energy equal to:

- 16.000.000 barrels of crude oil, which equals a fleet of 350 tankers – or
- the lifetime fuel consumption of 75.000 car owners – or:
- the electricity generated by eight average size nuclear power plants – or:
- the energy needed for industrial and commercial activities of 1,8 millions employees in total.

– **Material:** The savings of materials by remanufacturing amounts worldwide to 14.000.000 tons a year. This quantity of raw materials is equal to

- a railway train with 230.000 full cars, occupying a track of 1650 miles in length.

Regarding energy savings achieved by remanufacturing already at today's product ranges and volumes one can really state "A Star is Born". Imagine the energy savings that would be possible if remanufacturing activities are encouraged and continue their spread and growth in many more applications. This bright star might well be raised to a brilliant little galaxy giving light and life to nature and the environment on our planet and become just as indispensable as the sun shining every day.

Remanufacturing's New Horizons:
An Outlook to the Future

The approach of the magic year 2000 triggered new considerations about inspiring the proliferation of save-the-planet movements and focusing our minds towards a sustainable future in the new millenium.

The term "sustainable development" is understood as the principle that current economic progress should not endanger the prospect of future generations. Questions risen are world population growth, nutrition, employment and education, natural resource depletion, energy consumption, green house gas emissions, climate change and global warming.

Think Global - Act Local

The questions are of global scale. Answers however have to be found on local levels – in our existing industries and everyday lives.

Remanufacturing in its existing application and scale has presented itself as a proven solution generating multiple lives for our products; saving energy, natural resources, landfill requirement and reducing air / water pollution.

New Businesses and Jobs

Remanufacturing is creating new businesses and jobs at the same time – it has the potential to compensate or even turn around the losses of jobs in manufacturing employment.

Thus, in the scientific community, remanufacturing has already become recognized as a win-win situation for everyone, harmonizing economy and ecology. Remanufacturing represents an already existing solution, nut just a scientific approach to the "Factor 4" (double comfort / half resource depletion) discussion. The more, the better: Regarding product output / energy input, remanufacturing already matches manufacturing by "Factor 10".

From "Factor 4" to "Factor 10"

This good news will spread. University teachers as well as government officials have started to further promote the unique idea of remanufacturing. Lecturers are bringing remanufacturing into a discipline in engineering schools where they will also teach product design for recycling and remanufacturing.

Universities and Governments

Remanufacturing is appearing on the political agenda because it creates jobs and protects the environment.

Public administrations' and private companies' purchasing policies will prefer remanufacturable or remanufactured products when they buy office furniture, computer systems or components for their vehicle fleets.

Already today world class companies don't just buy remanufactured products – they also discover remanufacturing to boost their own productivity and competitiveness. Remanufacturing is also a business for the small, family-owned, local companies, which are the backbone of every national economy. These may tie the most intelligent knots in the global players' networks by remanufacturing.

World Class Companies

Eco-Innovation	Remanufacturing is an eco-innovation driver, with potentials on the economic and the ecologic sides as well. It will conquer new disciplines, new product areas and new markets – if public recognition also follows.
Bridging Technology and Society	We must not forget that the strongest driving force in our market place is always the consumer – technological "push" needs market "pull". Remanufacturing technology matters, but not as much as the people who will drive it and ultimately benefit. Fortunately, customer research indicates a rising awareness which is more than just lip service towards protecting the planet; particularly if consumers can have some fun and save money at the same time. Remanufacturing offers this magic twin opportunity.
	There is a strong potential for growth – the kind of healthy, balanced growth we need.
Take Action	Remanufacturers are in business at the right time in the history of this world to provide answers to many of our economic, environmental and employment challenges. Grasp the opportunity and take action.

Acknowledgements

This book contains 100 photos and graphics. A wide range of companies and institutions has made possible these illustrations, either by allowing the author to take photographs on their premises or by delivering pictures, information and data about their remanufacturing activities for publishing.

The 100 figures span two decades of remanufacturing and activities from three continents.
The publishers and the author have to thank the following companies, organisations and institutions, listed in alphabetical order, for their contributions (no. of figures in brackets):

All State Remanufacturing, USA (73, 74); **A**utomotive Parts Rebuilders Association APRA, USA and Europe (1, 30); **A**utomotive Rebuilders Supply, USA (28, 29);
BMW Bayerische Motoren Werke, Germany (36); **B**udweg, Denmark (48);
Champion Parts Rebuilders, USA (39); **C**overtronic, Germany (2, 79, 82); **C**ycle AK c/o VDMA, Germany (14);
Davies Office Refurbishing, USA (83, 84, 85, 86, 87, 88, 90); **D**eutz Service, Germany (42, 44, 50, 53, 57, 58); **D**irk Ihle, Germany (4, 5); DMT / HP, Germany (94);
Euro Motoren, Netherlands (31, 32, 33, 34, 60); **E**urope Auto Industrie, France (54, 61);
Fraunhofer Demonstration Center Product Cycles (1, 6, 13, 14, 21, 23, 24, 26, 30, 35, 38, 40, 41, 49, 56, 91, 92, 93, 95, 96, 97, 98, 99, 100), **F**ord Motor Company, USA (3);
Georgsmarienhütte, Germany (9, 10, 11)
Hewlett Packard / DMT, Germany (94); Hornberger Consulting, Germany (43);
Krueger Industries, USA, Mexico and Germany (45, 80, 81);
Linde, Germany (25); **L**ucas, Great Britain and Germany (Cover, 65, 66, 67, 68, 69, 89); **L**UK, Germany and Great Britain (37, 41, 46, 52, 62, 63, 64)
Meyer Bremen, Germany (7, 12, 20); **M**IP Mainz, Germany (15, 22, 27); **M**irec, Netherlands (18)
Oki, Germany (2)
Philips, Netherlands (19); **P**recision Starters and Alternators, USA (Cover, Inner Title, 51);
Remanufacturing Industries Council International RICI, USA (1); **R**ethmann Kreislaufwirtschaft, Germany (16, 17); **R**KW, Germany (56, 95);
Shredderwerke Herbertingen, Germany (8); **S**iemens-Nixdorf, Germany (76, 77), **S**ony Europe, Japan (78);
Unicardan Remanufacturing, Germany (2, 47, 55); **U**nidentified (59, final graphic); **V**erein Deutscher Ingenieure, Germany (24); **V**ereinigte Kunststoffwerke, Germany (25);
Westinghouse-Unimation, USA (70, 71, 72);
Xerox, Netherlands (75).

The above information about companies and countries are for photo source identification purposes only. They therefore do not necessarily represent the actual names and legal forms of companies or institutions mentioned. Due to continuous technological progress and market developments, also current remanufacturing activities of a particular company can have changed to a new technology or country of operation than the one mentioned in a particular photo.

Further Information about Remanufacturing

For further information about remanufacturing, the organisations and institutions who have supported this book can be contacted as follows.

Industry's Associations

Remanufacturing Industries Council International (RICI)
Contact: Bill Gager, Scott Parker
4401 Fair Lakes Court, Suite 210
Fairfax, Virginia 22033, USA
703 968-2995
http://www.remanufacturing.org

Automotive Parts Rebuilders Association
USA Contact: Bill Gager, Scott Parker
4401 Fair Lakes Court, Suite 210
Fairfax, Virginia 22033, USA
703 968-2772
European Division
Contact: Fernand J. Weiland
Forsbachstr. 13
D 51145 Köln, Germany
Tel.: xx49-2361/912-291
Fax: xx49-2361/912-791

Universities, R & D Centers

Boston University
Manufacturing Engineering Department
Professor Robert T. Lund
15, St. Mary's Street
Boston, Massachussetts 02215, USA

Rochester Institute of Technology
National Center for Remanufacturing
and Resource Recovery
Contact: Dr. Nabil Nasr
81 Lomb Memorial Drive
Rochester, New York 14623-5603, USA
716-475-6091
http://www-reman-rit.edu

Fraunhofer Demonstration Center
"Product Cycles"
Contact: Dr. Rolf Steinhilper, Martin Hieber
Nobelstr. 12
D 70569 Stuttgart, Germany
Tel.: xx49-711/970-1214 or -1116
Fax: xx49-711/970-1009
E-mail: ris @ipa.fhg.de; mbh @ipa.fhg.de